Hear this!

Sally Hewitt

QEB Publishing, Inc

First published in the United States by
QEB Publishing, Inc.
23062 La Cadena Drive
Laguna Hills, CA 92653

www.qeb-publishing.com

Library of Congress Control Number:
2005921170

ISBN 1-59566-125-5

Written by Sally Hewitt
Series Consultant Sally Morgan

Project Editor: Honor Head
Series designer: Zeta Jones
Photographer: Michael Wicks
Picture Researcher: Nic Dean

Publisher Steve Evans
Creative Director Louise Morley
Editorial Manager Jean Coppendale

Printed and bound in China

Picture credits

Key: t = top, b = bottom, m = middle,
c = centre, l = left, r = right

Corbis/Alain Nogues 8t, /Tom Stewart 12b, /Pat Doyle 14l, /Norbert Schaefer 14r, / ROB & SAS 15, /Jim Craigmyle 18;

Getty Images/Elyse Lewin/Photographers Choice 4, /Chris Windsor/The Image Bank 5b, /Richard Price/Taxi 5t, /Romily Lockyer/The Image Bank 6b, /Macchina/Stone 8b, /David Tipling/Photographers Choice 9t.

The words in bold **like this** are explained in the Glossary on page 22.

Contents

Hear this

You have five senses that give you all kinds of information about what is going on around you.

The five senses are sight, touch, taste, smell, and **hearing**.

This book is about hearing.

◀ Your sense of hearing helps you to understand what people say.

Some sounds warn you of danger. A speeding ambulance has a **loud** siren. It means—

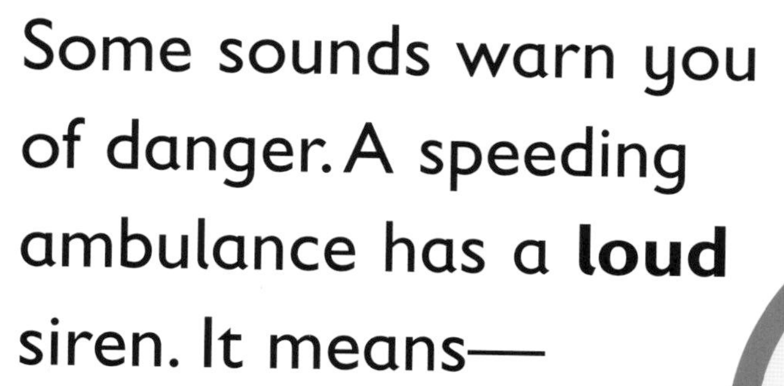

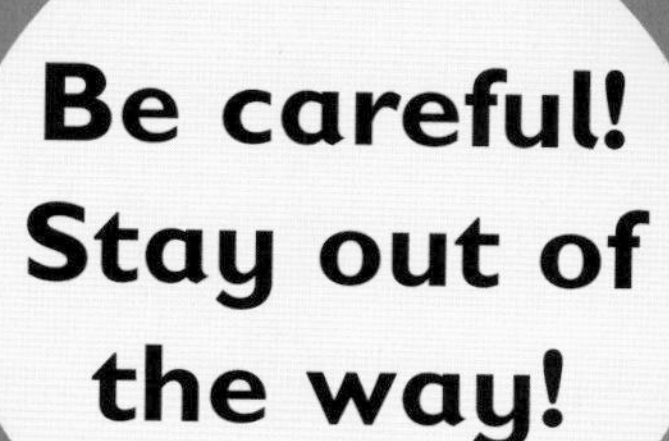

Be careful! Stay out of the way!

Music is a sound that you can enjoy. Do you like listening to music?

What other sounds do you like to hear?

Listen!

The world around you is full of sounds. What can you hear now?

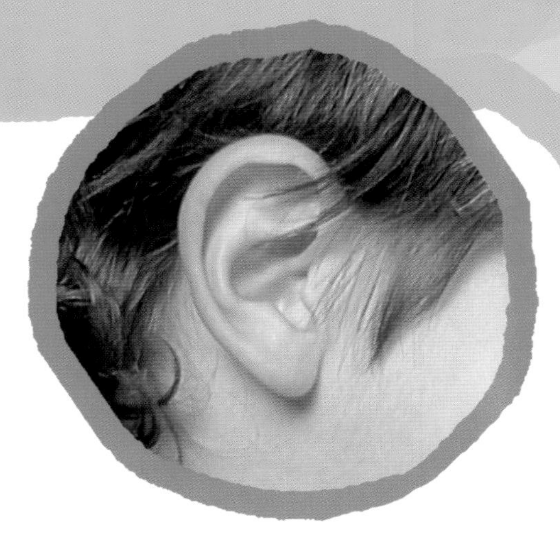

▲ Your **ears** are the part of your body you use to hear.

Listen! Can you hear any sounds you've never heard before?

You can hear air moving inside a shell. Does it sound like the ocean?

You sometimes know what is happening around you by listening to the noises.

Activity

Ask a friend to shut their eyes and listen while you:

- bounce a ball
- pour milk into a glass
- tear a piece of paper
- bite into an apple

Can your friend tell you what you're doing by the sounds you make?

Sound waves

You can hear the sound of helicopter blades turning.

The blades move the air around them and make **sound waves**.

You can't see sound waves, but you can hear them.

A bird singing makes a different pattern of sound waves.

You hear when sound waves go into your ears. Your ears send a message to your brain. Your brain tells you what you are hearing.

Where's that sound?

Your ears are small and flat. You have to turn your head to move them.

◂ Cup your hand around your ear and point it toward a sound. Does it sound louder now?

Different animals have differently shaped ears. Some animals can hear better than you.

▸ This dog can turn its ears toward a sound without moving its head.

You have two ears to help you find out where a sound is coming from. You can usually tell where someone is by listening to the sounds they make.

Activity

Ask a friend to shut his or her eyes. Tiptoe around them, then stop and make a tiny squeak.

Can your friend point to where you are just by listening?

Eardrums

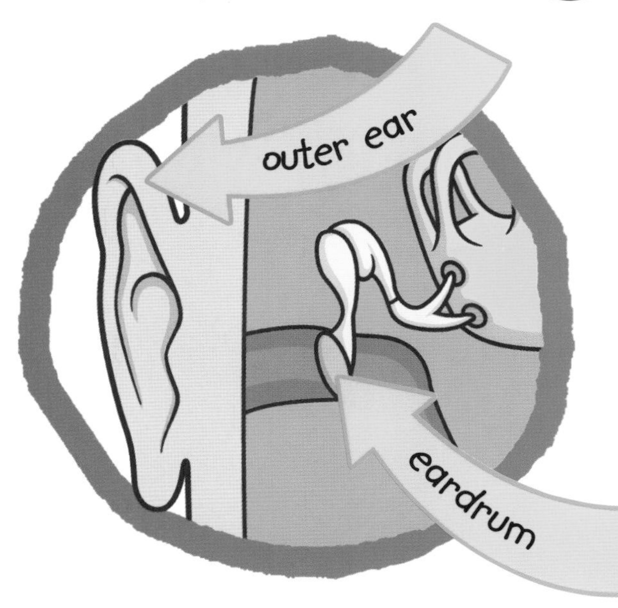

Your **eardrum** is a small piece of skin inside your ear. It is stretched tightly like a drum skin.

When you bang a drum, the skin **vibrates**. This means it pushes the air around it backward and forward.

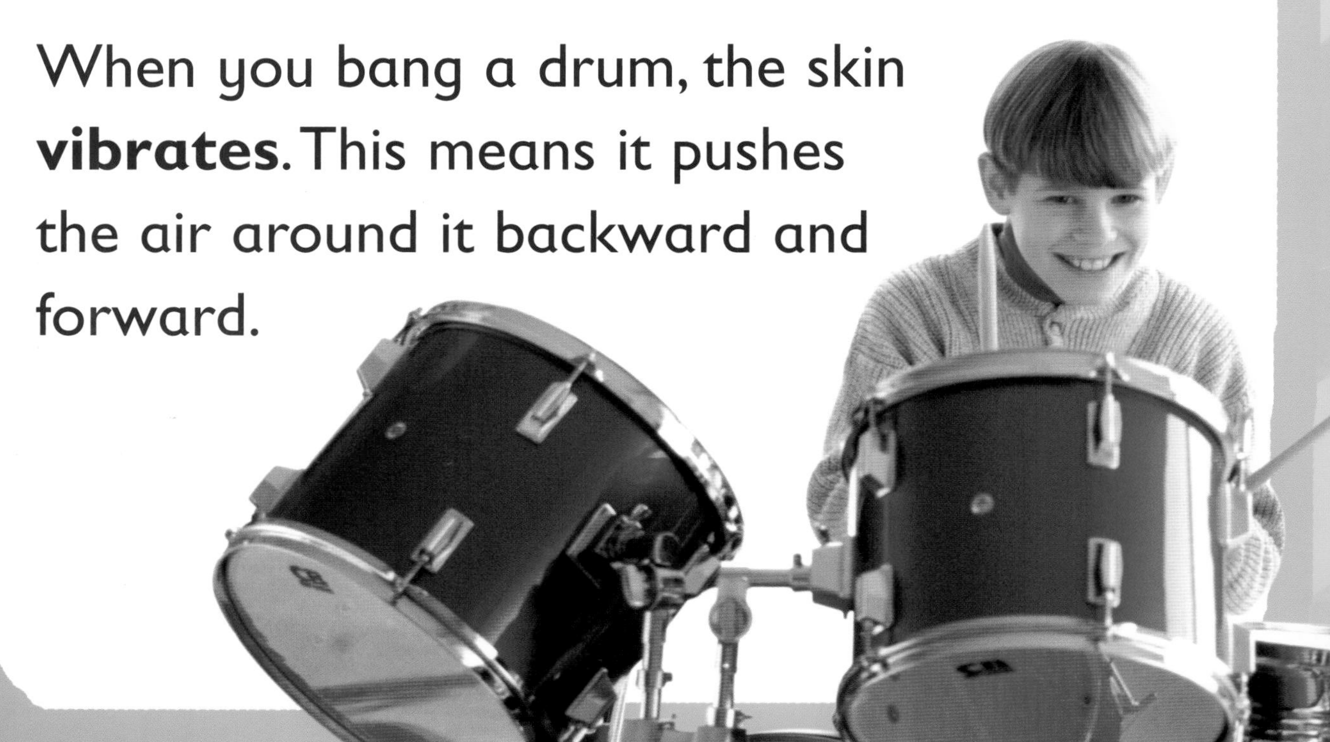

When a sound goes into your ears, it hits your eardrum and makes it vibrate.

Activity

Make your own sound waves.

- Stretch some plastic wrap over a bowl. (This is like your eardrum.)
- Sprinkle rice on top.
- Bang a baking pan with a wooden spoon just above it.
- Watch the sound waves vibrate the plastic wrap and make the rice jump.

Loud and quiet

Loud sounds make very large vibrations in the air. They can hurt your ears!

You can put your hands over your ears to keep out some of the sound.

A dog barking is a loud sound. What other loud sounds have you heard?

Quiet sounds make small vibrations.
You have to listen carefully to hear them.

You can hear **quiet** sounds better when they are close to your ear.

Whispering is a quiet sound.
What other quiet sounds have you heard?

Near and far

A car engine sounds loud when you are near it. The sound becomes quieter and quieter as the car moves farther away.

As sound travels through the air, it spreads out and becomes quieter.

Activity

Your voice can travel along string into your friend's ear.

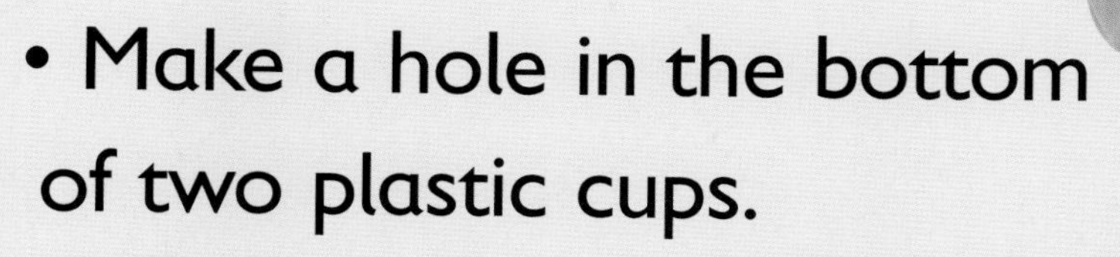

- Make a hole in the bottom of two plastic cups.
- Thread a long piece of string through each hole.
- Tie buttons at each end.
- Give one cup to a friend and pull the string tight.
- Talk into the other cup. Your friend will hear your voice at the other end.

High and low

Musical instruments play **high** and **low** notes.

The thick strings on a guitar play low notes.

The thin strings on a guitar play higher notes.

◂ The strings on a guitar play different notes.

Activity

You can play high and low notes on a glass of water.

- Pour yourself a glass of water or juice.
- Tap the glass gently with a spoon and listen to the sound.
- Take a sip. Tap the glass again.
- How has the sound changed?
- Tap every time you take a sip.

Does the note get higher or lower?

I can't hear

People who can't hear learn to tell what someone is saying by watching their face and lips.

Some deaf people can make signs with their hands to talk to each other. This is called sign language.

These children are saying "hello" in sign language.

You can mime, or make movements, to tell a story without speaking.

What story could this boy be telling?

Can you tell a story without saying a word?

Glossary

Eardrum
A small piece of skin inside your ear like a drum.

Ears
You have two ears. They are the parts of your body you hear with.

Hearing
This is one of your five senses. You hear with your ears.

High
Sounds can be high. A bird singing makes a high sound.

Loud
Sounds can be loud. A big truck makes a loud sound.

Low
Sounds can be low. A dog growling makes a low sound.

Quiet
Sounds can be quiet. Whispering is a quiet sound.

Sound waves
Noise travels through the air in waves called sound waves.

Vibrate
To move backward and forward a tiny amount very fast.

Index

Parents' and teachers' notes

- Look throughout the book for words about sound, such as loud, quiet, high, and low. Listen to different sounds. Use these words to talk about the sounds you hear.

- Make a tape of everyday sounds, play them back, and ask your child to guess the sound. Find new words to talk about the sounds, such as splash, squeak, bang, patter, and scrape.

- You can feel vibrations. Show your child how to put his or her hand on the speaker of a radio or CD player to feel vibrations as the sound comes through.

- Make a big cardboard cone and use it to magnify sound. Put the end of the cone around, not in, your child's ear. Point the cone toward a quiet sound. Does it make the sound louder?

- Cut out pictures of animals. Discuss the different sounds they make.

- Find pictures of things that make a noise, such as a leaf falling, a watch ticking, people whispering, someone shouting, a train, and an airplane taking off. Ask your child to put them in order, starting with the quietest.

- Look at pages 12 and 13. Talk about why loud noises, such as shouting, screaming, banging, and drilling, can hurt your ears. Find out which workers protect their ears from loud noise when they work.

- When you walk along a busy street, discuss the sounds of the traffic. Which ones are the loudest vehicles? Which are the quietest? Notice how the sound of traffic gets louder as it approaches and quieter as it goes away from you.

- Ask your child to draw a picture of their face smiling, surrounded by pictures of sounds they like. They can draw another picture of their face frowning, surrounded by pictures of sounds they don't like. Talk about why they like some sounds and dislike others.

- Collect small containers with lids. Put rice, dried pasta and beans, and coins inside. Shake the containers and listen to the sounds they make.

- Look at pictures of different places, such as the beach, an airport, or a farm. Talk about the sounds you hear at these places.